LET'S-READ-AND-FIND-OUT BOOK CLUB

A JELLYFISH IS NOT A FISH

by John F. Waters

Illustrated by Kazue Mizumura

Thomas Y. Crowell New York

Jellyfish swim in cold oceans and warm seas.
They swim along the shore and in deep water.
Some jellyfish live in lakes and ponds.

You may have seen jellyfish in the water where you swim. Some jellyfish are harmless. You can even touch them. People scoop them out of the water and hold them in their hands.

But some jellyfish can sting. And a few jellyfish are poisonous. So don't get near them or try to pick them up unless you know which ones are safe.

A jellyfish is not a fish at all. It doesn't even look like a fish. It has no head, no bones, no scales, and no fins. It looks like a bell or an open umbrella. Tentacles hang down from the

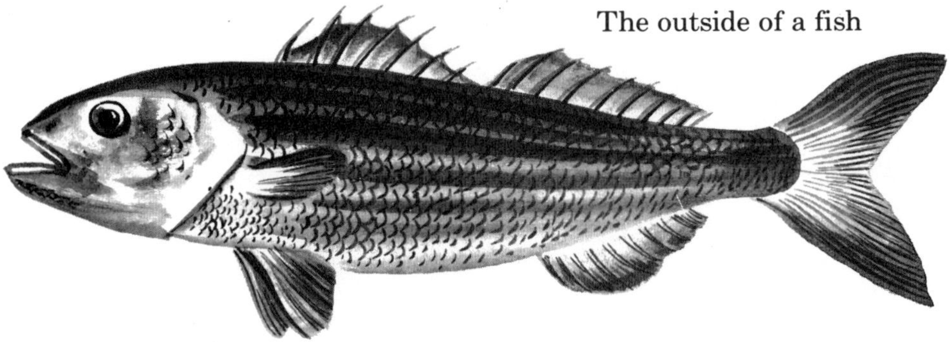

The outside of a fish

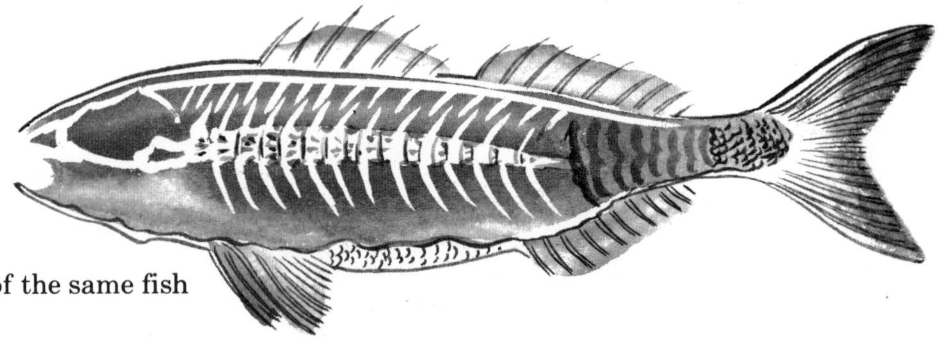

The inside of the same fish

edge of the umbrella. The tentacles move up and down and around like loose strings. Some people think the tentacles look like skinny snakes.

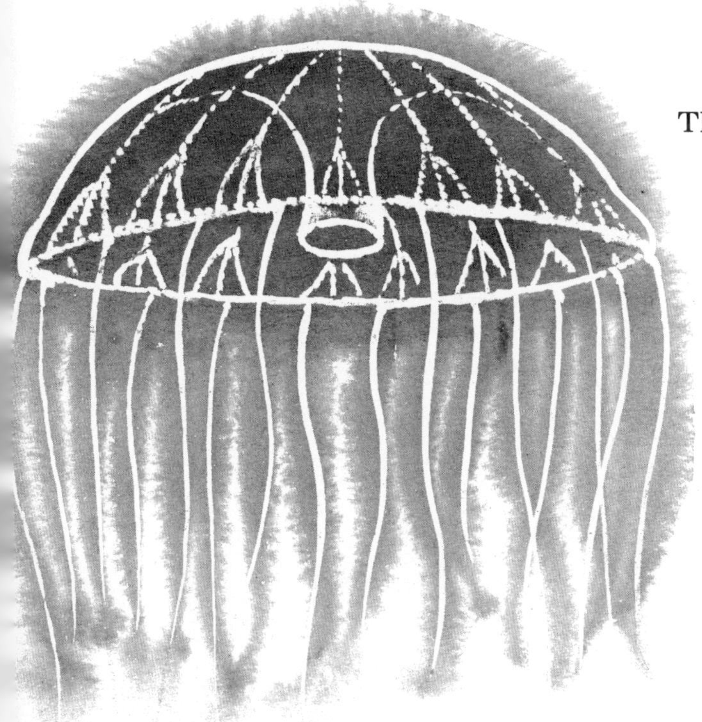

The outside of a jellyfish

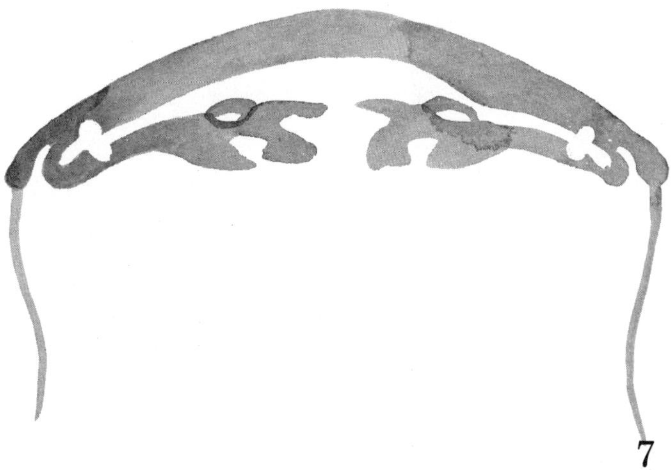

The inside of the same jellyfish

All jellyfish belong to a big group of animals. One name for the group is medusa. Jellyfish are called this because their tentacles remind people of a Greek goddess named Medusa. Her head was covered with snakes instead of hair.

There are about three thousand kinds of jellyfish. Some are so tiny you can hardly see them. Twenty-five of them side by side would cover only an inch on a ruler. Big jellyfish may be as big as a barrel or a double bed. They weigh hundreds of pounds. These giant jellyfish live in the cold waters of the far north and the far south.

The moon jellyfish lives just about everywhere. It is found in all the seas of the world. It can be yellow, pink, or blue. It is soft and round and has a wavy edge. Some moon jellyfish are very small, only an inch across. Others are bigger, five or six inches across.

The moon jellyfish has short tentacles. Other kinds of jellyfish have long ones. Some tentacles are as long as a tall man, even longer.

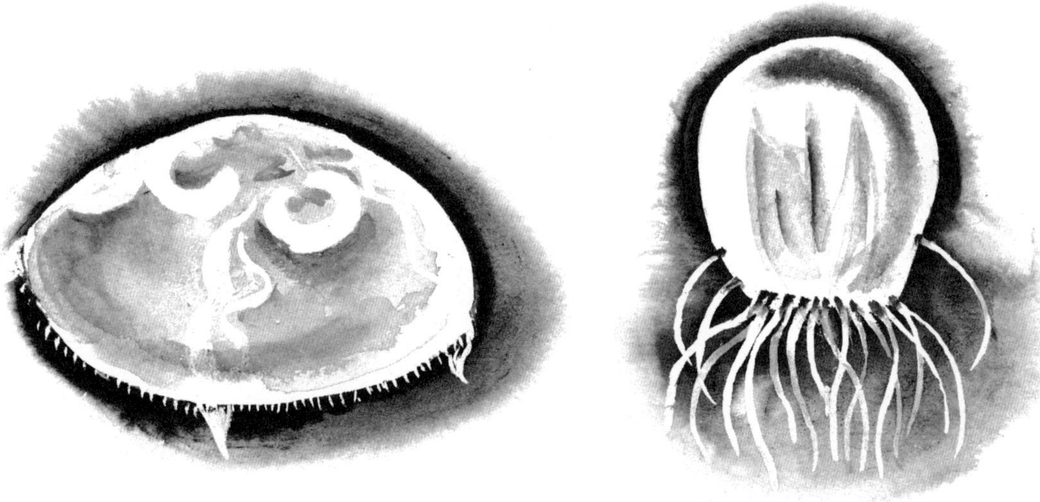

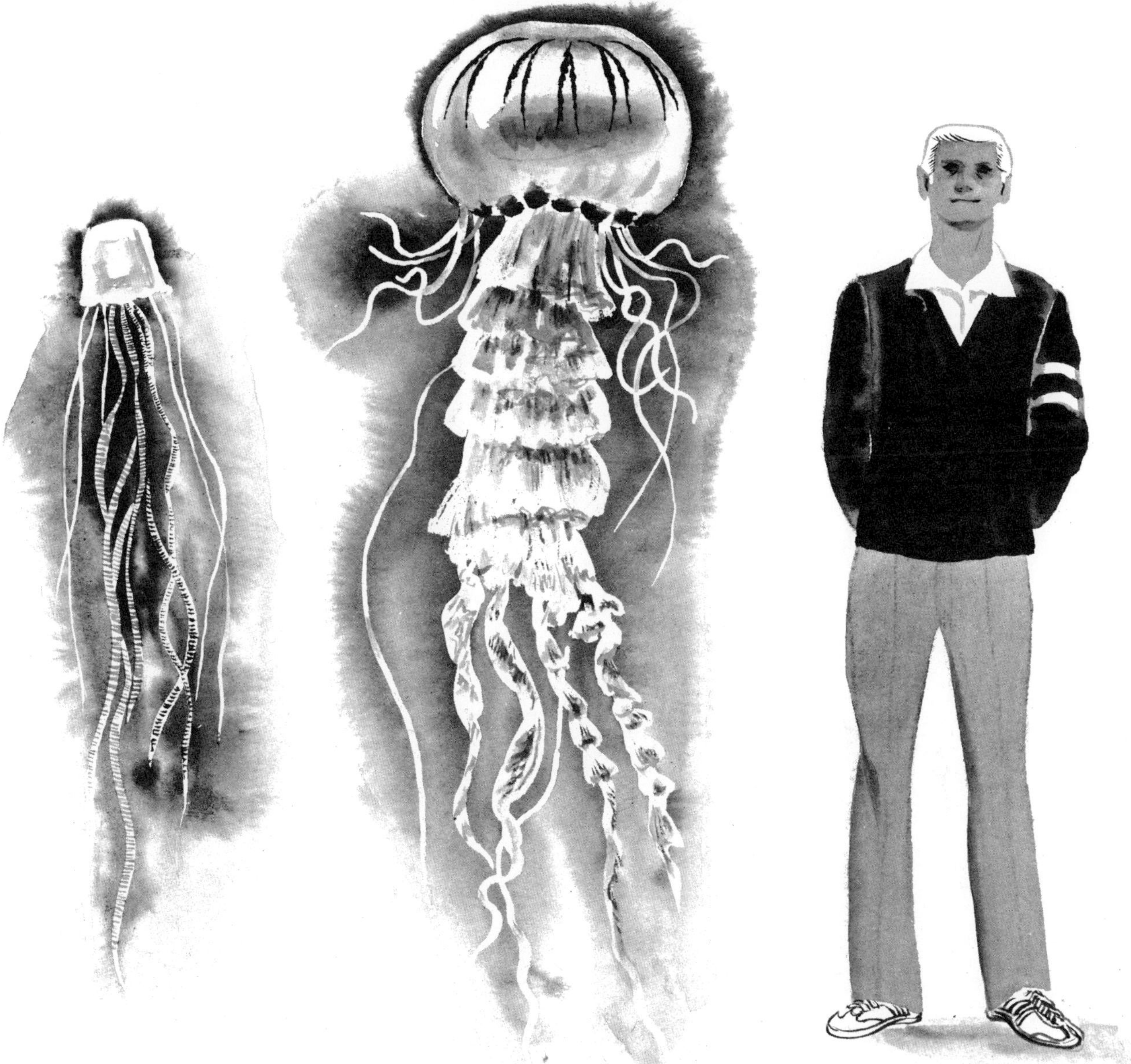

Most jellyfish have stingers on their tentacles. When the tentacles touch a fish, the stingers go into the fish. They poison the fish. The poisoned fish can't swim away, and so the jellyfish can eat it. Sometimes the fish is as big as the jellyfish.

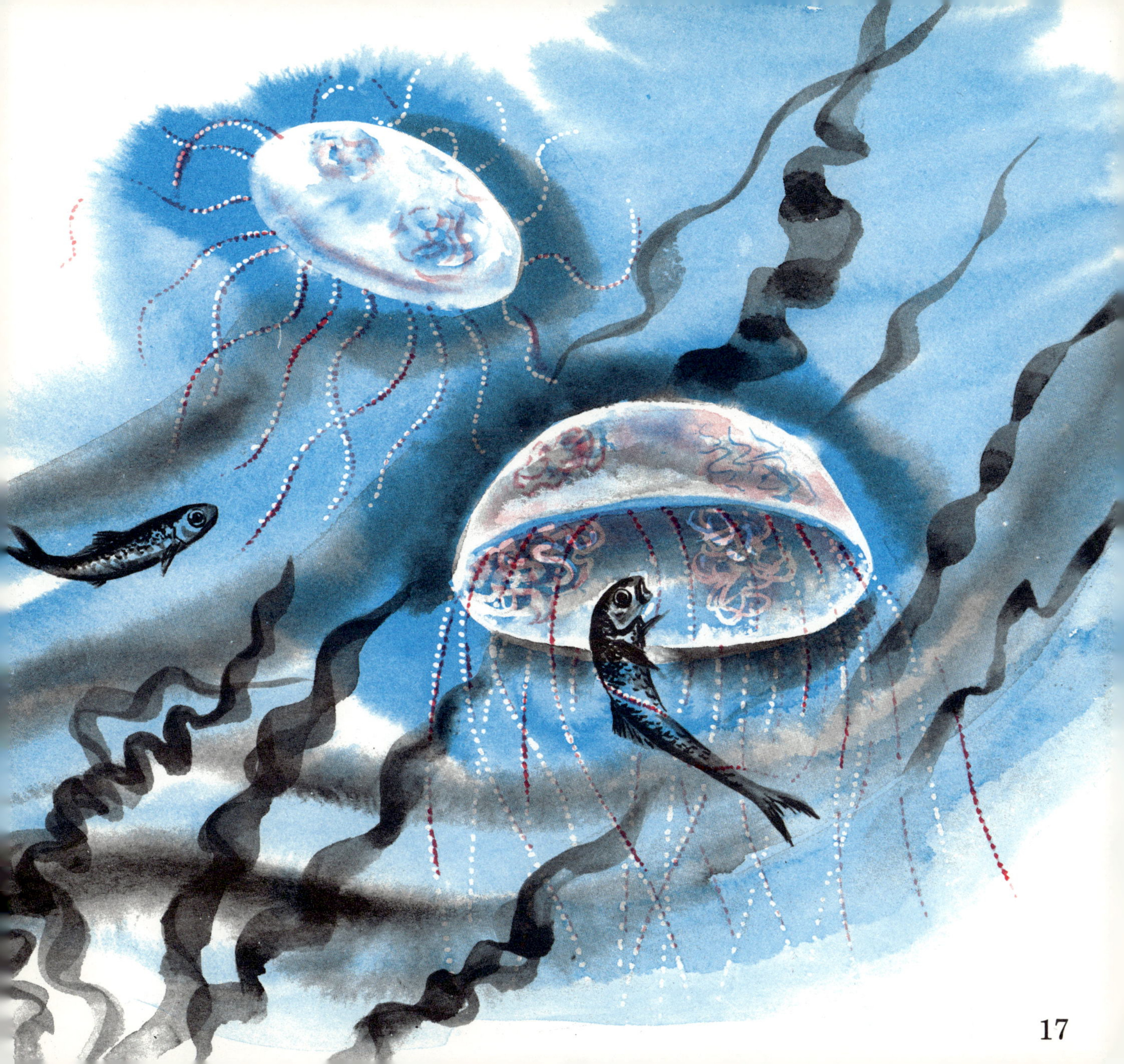

A dangerous kind of jellyfish is called the Portuguese man-of-war. It lives in the warm waters of the world. In summer in the Atlantic Ocean it sometimes drifts as far north as Cape Cod. This jellyfish is sky blue, pink, and purple. Its long tentacles hang from an odd-shaped bag or float. The float looks something like the helmet that Portuguese soldiers used to wear. That is how this jellyfish got its name.

The Portuguese man-of-war is unusual in many ways. It looks and acts like a single animal. But it is really many animals joined together. It is a whole colony of animals. Some of the animals in the group sting food and catch it. Others digest the food. Still others lay eggs and produce new animals for the colony.

Swimmers can be poisoned by a Portuguese man-of-war. So can fish. But there is at least one kind of fish that is usually not hurt by the poison. It is called the man-of-war fish. It lives among the tentacles, eating scraps of food left over from the jellyfish's meal.

Other jellyfish are even more dangerous. The most deadly jellyfish is called the sea wasp. Some sea wasps are found along the north coast of Australia. Every year people die from their stings. In twenty-five years sixty people were killed by sea wasps. In the same time only thirteen people were killed by sharks. In Australia, sea wasps are more dangerous to swimmers than sharks.

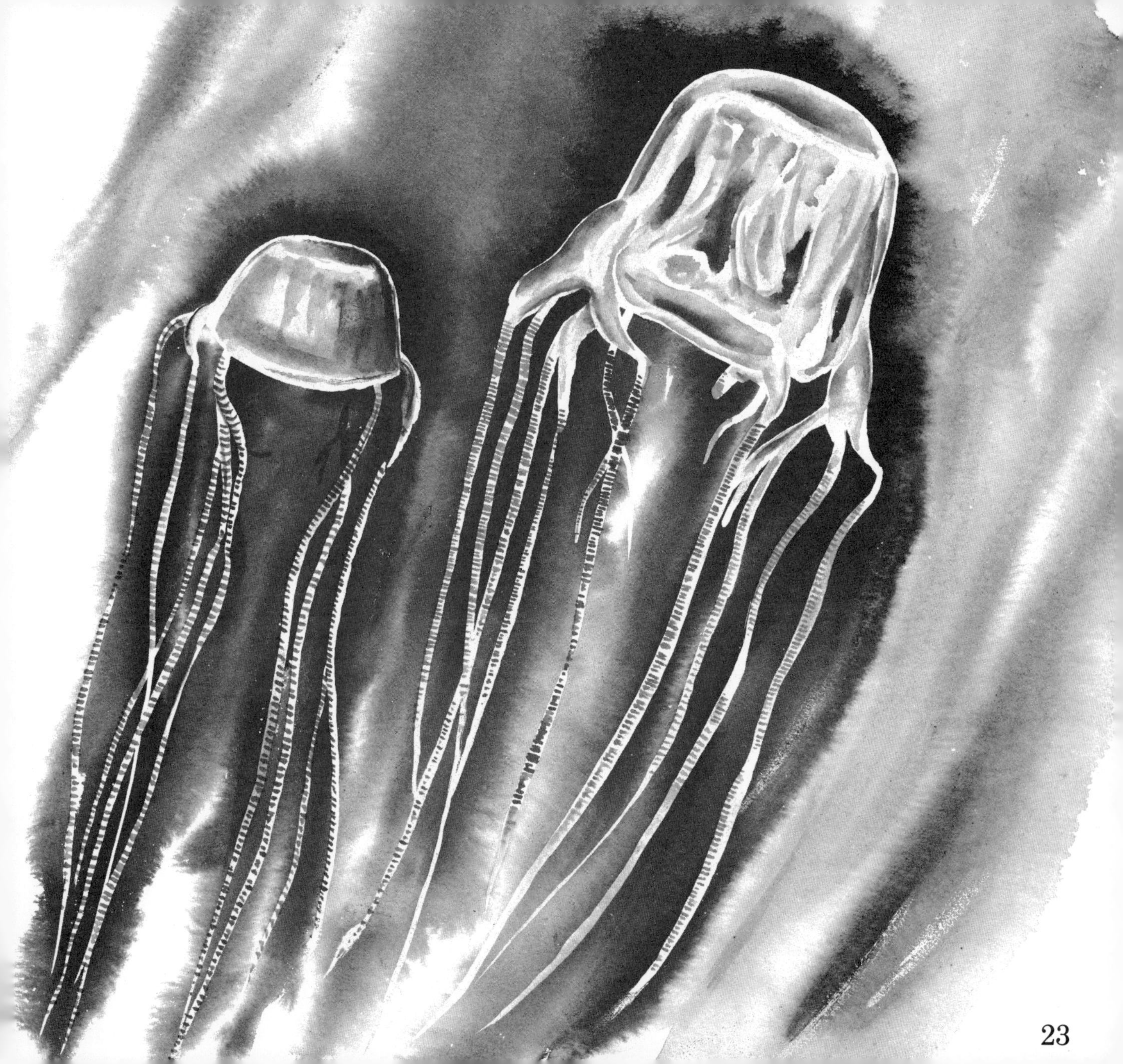

A jellyfish is like a floating bag of jelly. The jelly is about as stiff as the gelatin you may eat for dessert. On the underside of the bag is a single opening. It is the mouth of the jellyfish. Food goes in the mouth and waste comes out. Jellyfish eat fish, shrimp, crabs, and other jellyfish. They also eat tiny plants and animals that drift in the sea.

Jellyfish are simple animals, much simpler than fish. Jellyfish have very few parts, other than their big gaping mouths. Many of them don't even have eyes. Those that do can't see things clearly. They can only tell light from darkness.

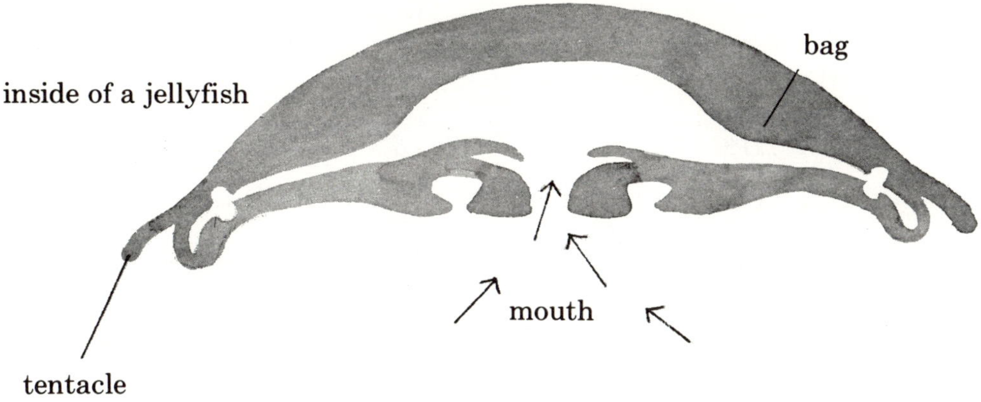

The inside of a jellyfish

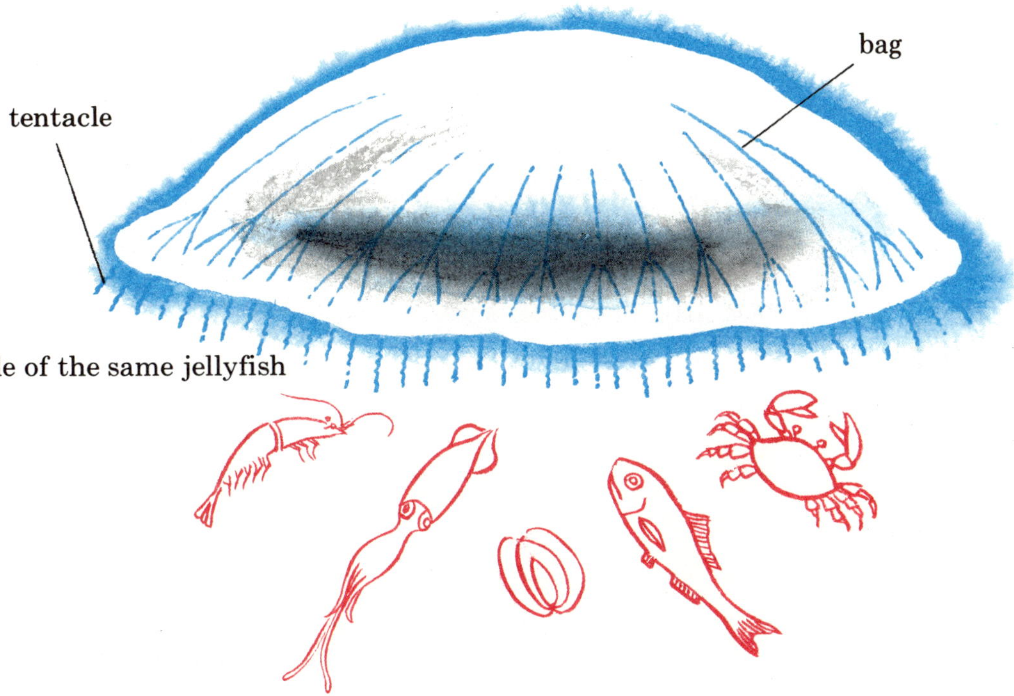

The outside of the same jellyfish

Jellyfish move by squirting water out of their mouths. The bell of a jellyfish moves in and out, in and out. When the bell moves out, water

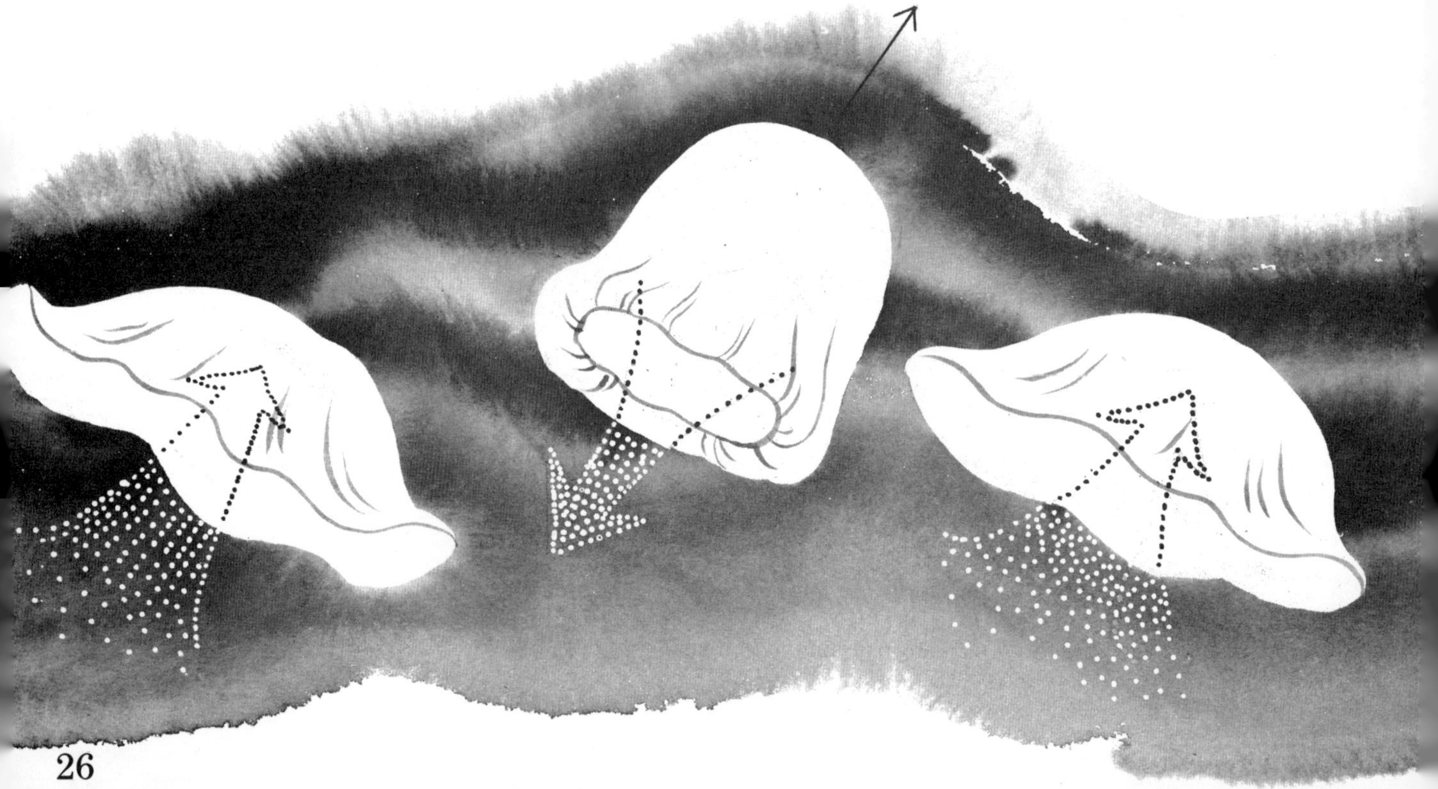

rushes into the mouth of the jellyfish. When the bell moves in, the water jets out of the mouth. This makes the jellyfish move ahead.

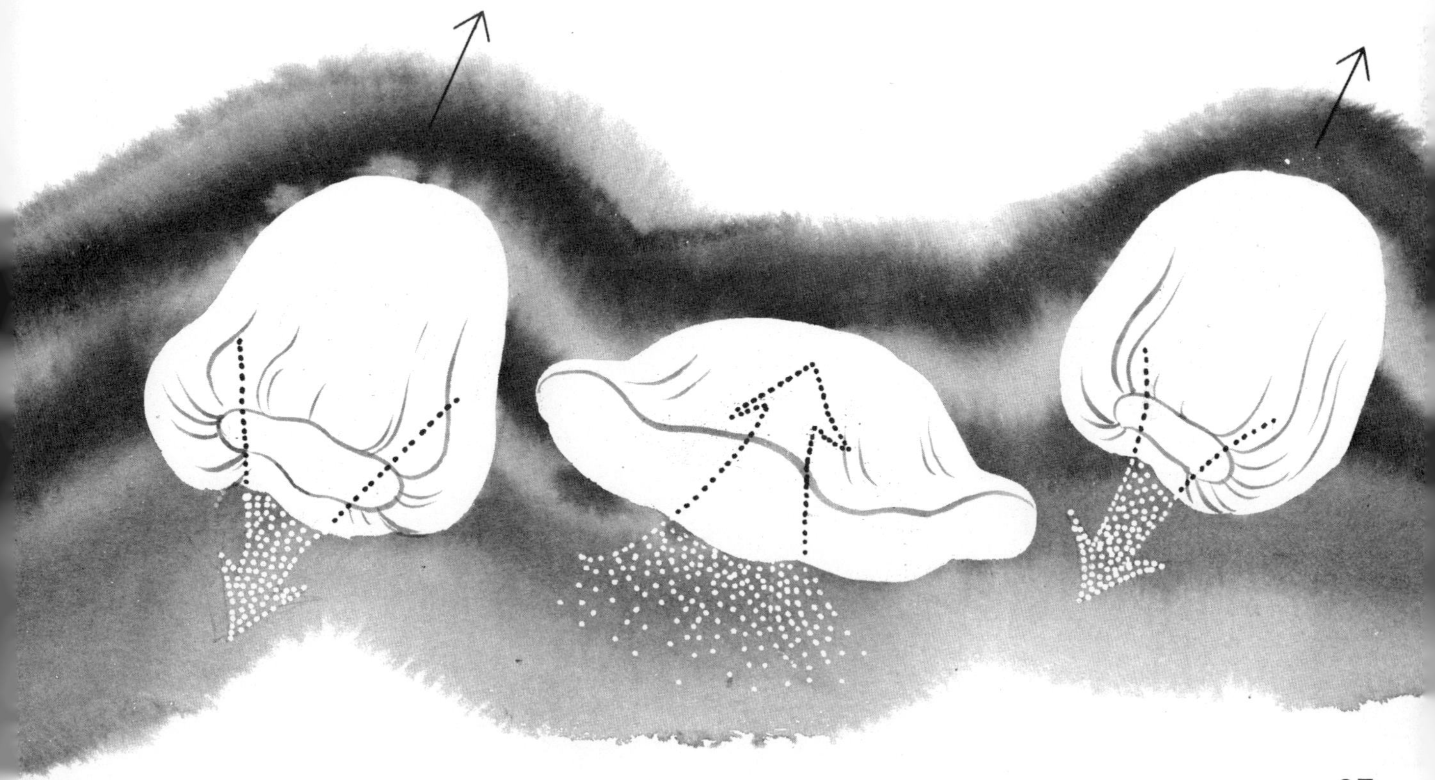

Some jellyfish glow with a pale blue or green light. The light goes on and off every few seconds. No one knows why these jellyfish turn on and off like this. Maybe the light attracts fish, so the jellyfish can sting them and eat them. No one knows for sure.

Jellyfish are in all the oceans of the world. And they have been there for a long, long time, millions and millions of years. The first jellyfish were swimming in the oceans before dinosaurs lived on land—even before any real fish had appeared in the seas.

The next time you see a jellyfish in the water, watch it to see how it swims and how the tentacles swing about like the snakes on the head of Medusa. But don't touch it unless you are sure it won't sting you. Don't even touch a jellyfish that has washed up onto the beach. The jellyfish is probably dead, but its stingers can still hurt you.